Providing Safe, Healthy, and Functional WorkSpaces
Fred Fanning

Copyright © 2015 Fred Fanning
Library of Congress Control Number: 2015919654
CreateSpace Independent Publishing Platform, North Charleston, SC
BISAC: Technology & Engineering/Construction/General
ISBN-13: 978-1512232288
ISBN-10: 1512232289

This publication is designed to provide accurate and authoritative information concerning the subject matter covered. It is sold with the understanding that the author is not engaged in rendering legal, accounting, or other professional services. If legal advice or other expert assistance is required, the services of a competent professional person should be sought

The reader should not rely on this publication to address specific questions that apply to a particular set of facts. The author makes no representation or warranty, expressed or implied, as to the completeness, correctness, or utility of the information in this publication. The author assumes no liability of any kind resulting from the use of or reliance upon the contents of this book.

# Dedication

I dedicated this book to the men and women of the Logistics and Facility Management Office of the Headquarters, Department of Energy. The hard work done daily by Michael Shincovich, Cherylynne Williams, Michael Wolfe, Lisa Peteet, Ed Danchik, Clint Cleveland, Michael Watkins, Terry Butler, Ronell Nichols, Jerry Vann, Jesse Stephens, Ashley Jessen, Donna Blumenauer, Melissa Edmonds, and Andi James provides thousands a safe, healthy, and productive workplace.

# Contents

# Introduction

I worked in facilities and logistics for many years. During that time, I learned a lot of valuable lessons about managing and altering space inside buildings that I would like to share. When I started in facility management, I looked for a single book I could use to help me understand the process. I was not able to find such a book. I did find resources that I reference in this book, and I encourage you to read them for yourself.

The International Facility Management Association (Facility Management, 2013) defines Facility Management as the practice of coordinating the physical workplace with the people and work of the organization. There are several divisions within Facility Management that each fulfill a part of the Facility Mission. Space Management is usually one of the divisions. This division manages the acquisition and allocation of space within a facility to various users of an organization or as a landlord renting space to tenants. This means that the personnel within this division must know real estate, space allocation, asset management, facility alterations, furniture design and procurement, flooring installation, contracting, and project management.

When a building is first built the exterior and interior are usually designed by a large design firm and constructed by a large construction contractor. The building is normally delivered to the new owner in completed condition. Unfortunately, over the years organizations change or tenants change, and before you know it, the interior of the building needs to be changed or perhaps a new or another building is needed. That is where the Space Management staff comes in. If additional space or newly leased space is to be acquired, the Space Management staff also performs those duties. The Space Management staff will not design and build a new facility. For a newly leased facility, they work with the landlord to design and build the facility. Finally, the Space Management staff provides all the furniture, fixtures, and equipment needed to make interior spaces work. They serve as a one-stop shop for all space needs. In some organizations, the Space Management staff will also acquire and install telephony and IT systems. Over the years, separate Information Management sections have evolved to do all telephony and IT systems work.

The Space Management staff has two main customers. First is the person who has space and wants it changed or wants more or less space, and

1

second is the facility operations staff. It is one thing to meet the user's needs, but quite another to meet the building maintainer's needs. Space managers must build out space that can be maintained. This means that sometimes the customer cannot have exactly what he or she wants because of the issues involved with the building systems and maintenance efforts. The Space Management staff walks a fine line to get both of these customers what they want.

The Space Management staff I refer to can be employees of the Facility Management office, contractors, or a mixture of both. The trend now is for a few Facility Management staff to be employees and the rest of the workforce to be contracted. In my opinion that seems to work fine; however, this is not a one size fits all solution. The office should be staffed with the right people to get the job done.

The individual in charge of the Space Management Division is usually referred to as the Space Manager or Space and Alterations Manager. This individual is usually a senior facility management professional. Space Management Division staff are Facility Management Professionals, Interior Designers, AutoCAD drafters, Architects, and Engineers. These workers are broken down into two groups of Space Planners and Facility Specialists.

Space Planners perform the following work:

- Lease space when required.
- Visit sites to verify space availability.
- Ensure proposed site or space is feasible for the space requirements.
- Conduct studies and surveys to analyze current and long-term space requirements.
- Provide alternate use of space if possible.
- Advise and recommend policies for the use of space.
- Submit quarterly and semi-annually two reports -- "Organizational Assignment by Locations" and "Organizational Personnel Assignment by Fiscal Year."
- Also, provide weekly "Vacant Space Report."
- Identify customer requirements for alterations to interior workspace.
- Develop AutoCAD floor plans that meet customer requirements.
- Develop procurement documents to contract work.
- Develop independent cost estimates.
- Work with procurement staff to contract work.

## Providing Safe, Healthy, and Functional Workspaces

Most Facility Specialists work in the operations and maintenance side of Facility Management. Facility Specialist performs the following work:

- Request Space Planner review of contractor's bids that are over 30% higher than the independent cost estimate.
- Coordinates with Procurement staff to conduct a walkthrough with potential contractors.
- Provide oversight of all contract work.
- Verifies work is completed.
- Conducts punch list and ensures all issues are corrected.
- Obtain customer's acceptance of work.
- Performs final billing and charge back to customers.

Let me share the bad news with you right up front. Managing or working in space is a thankless job. You will likely never make customers completely happy. Everyone seems to want more, now, and at less cost. For that reason, the job involves compromise and the ability to negotiate with what I call stakeholders to get the best outcome for everyone. I use the word stakeholders to identify anyone who has a stake in or will be affected by the effort undertaken by the Space Management Division, including:

- Employees
- Customers
- Facility Maintenance Personnel

For more information on this and other topics, please visit the bibliography at the end of this book.

# Chapter 1 – Basics of Space Management

## Introduction

It is important for everyone involved in space management to be familiar with the basics. In this chapter, I will provide familiarization with space measurement, allocation, and provide information I learned the hard way. This will help explain how space is managed effectively. In space management, you will hear the terms real property and real estate. The terms mean the same thing and are used interchangeably. You will also hear the following terms used: gross square (Sq.) feet (Ft.), rentable Sq. Ft., and usable Sq. Ft. Not understanding what each of these terms mean can cost thousands of dollars in overpayments. Here is a definition of each:

- Building Gross Area – the largest area of building measurement that includes interior gross area, voids and interstitial space, and exterior wall thickness.
- Interior Gross Area – the largest measurement of building interior space that includes occupant area, occupant storage, building and floor amenity area, building and floor service areas, base building circulation, major vertical penetrations, and interior parking.
- Rentable Sq. Ft. – the measurement of building an interior that you pay rent on that includes occupant area, occupant storage, building and floor service areas, building and floor amenity areas, and base building circulation.
- Usable Sq. Ft. – the smallest measurement of building an interior that includes the portion of a floor or building used to house personnel, equipment, fixtures, furniture, supplies, goods or merchandise, and secondary circulation.

It is essential that everyone understands these terms and measurements so that when agreements are made the results are what everyone expects.

## Space Measurement

"The physical dimensions of a facility or space are the special context within which the manager executes his responsibilities (Cotts, 2010). Measuring space is one of the first things you need to know. Space comes in Sq. Ft. This includes the width and length of space. There are several reasons for using a standard model (Space Management, 2013):

- Consistent with standard industry practice.
- Provides common terminology.
- Facilitates comparison with industry benchmarks.
- Supports decision-making.
- In combination with other data, provides visibility of cost, occupancy, utilization, and value.

I like the unified approach to measuring office space as outlined by the American National Standards Institute (ANSI)/Building Owners and Managers Association (BOMA) in Z65.1-2010, Office Buildings: Standard Methods of Measurement. This unified approach provides the following (Space Management, 2013):

- Facilitates clear communication among commercial real estate professionals.
- Fosters consistent, unambiguous measurement of rentable areas.
- Applies to both existing and new office buildings.

Space management guidelines must be developed for the organization that ensures space is designed and assigned based on function and use. The target utilization of 200 Sq. Ft. per person or less is used as an essential guide. This guideline also outlines several other responsibilities.

Space Planners maintain space allocation and organizational assignments. They identify all the space and who is using it through a process called Space Allocation and Assignment. This is done by meeting with organizations and walking the space to identify who is using what.

Surveys must be conducted of existing conditions. This helps to ensure that the space allocations (i.e., occupied and vacant space) are accurate and labeled with the correct official title. These surveys are coordinated with the organizational managers as necessary for space assignments.

The Space Planners will cross-reference the space occupied against the existing conditions to ensure correct official title, and square footage is represented in key plans. They then develop and transfer space allocation and assignment data into a space management database.

## Space Allocation and Tracking

A headcount is done annually to count every worker in a facility. These numbers are then used to determine updated utilization rates. The utilization rate is the number of Sq. Ft. of space divided by the number of people. If you have 10,000 Sq. Ft. of usable space and you put 60 people in that space, your utilization is 167 Sq. Ft. per person. Is that good or bad? There is no right answer to this question. My experience has been that a range of 170 to 200 Sq. Ft. per person provides a good work environment. This utilization rate is an all-in number that means it includes offices, cubicle rooms, file rooms, conference rooms, teaming rooms, break rooms, storage rooms, and interior hallways. There is another utilization rate that is office-only that is only office and cubicle space. The utilization rate for Office-Only should be between 115 and 135 Sq. Ft. per person. The utilization rate is an average that allows some workers to have 200 Sq. Ft. while other workers have 150 Sq. Ft. The difference might be for pay grade or seniority.

The information about space measurements in the organization should be documented in a table that reflects the different units within the organization. The table would look like the example in Table 1. As you can see in the table, the overall space utilization is 203 Sq. feet per person. Even though the CEO's office exceeds 200 Sq. Ft the other units use well below 200 Sq. Ft. This is how it will normally work. The space provided to a particular unit is based on the work completed by that unit. The Human Resources and Facility Management Offices use both offices and cubicles while the Sales Office uses cubicles. Within units, they can have conference and team rooms, but the Sq. Ft. count against that unit.

| Unit | Number of Employees | Gross Sq. Ft. | Rentable Sq. Ft. | Usable Sq. Ft. | Utilization Rate |
|---|---|---|---|---|---|
| CEO Office | 8 | 4,200 | 3,650 | 3,500 | 438 |
| Human Resources | 22 | 4,800 | 4,360 | 4,000 | 182 |
| Sales | 34 | 6,500 | 6,420 | 6,200 | 182 |
| Civil Rights | 4 | 650 | 620 | 500 | 167 |
| Facility Management | 8 | 1,500 | 1,450 | 1,200 | 150 |
| Total | 76 | 34,150 | 16,500 | 15,400 | 203 |

Table 1 – Organizational Space Measurement Table

## Space Analyses

Space Planners conduct space analyses to determine the most effective way to use space. These analyses include reviewing key plans and as-built drawings. Options are then developed to meet management requirements for space. Schematic drawings are prepared at 1/8" = 1'-0" scale. Presentation packages are prepared to consist of floor plans, elevations, finish selections, and perspectives as necessary for management decisions. The graphic representation is provided on key plans of proposed options, pro and con analyses, cost estimates, and proposed schedules. The presentation packages are then presented in a clear and concise manner. Revisions and multiple iterations may be required before management approval. Once management approves, the personnel are moved to the optimum locations that demonstrate the best use of space.

## Centralized or Decentralized Space Management

There are pros and cons to decentralizing the management of space. This method assigns space to the unit or office manager to control and pay rent for. The primary reason this is done is to better manage space. The rationale is that if a unit or office manager has control of space and pays for it, they will use it better, saving resources. I have never seen it work this way. What usually happens is that the unit or office manager hangs on to space as long as his budget supports it. When another unit or office needs space, the other office refuses to release it. The Space Manager has little ability to control space using this method, which leads to wasted space costing thousands of dollars. The focus of managing space is on the benefit of the unit or office.

When space management is centralized, the Space Manager controls it. The cost of space is borne by the Facility Management Office. When a unit or office manager needs space, the Space Manager directs moves that are for the good organization, not a particular unit or office. This method results in the most efficient use of resources saving thousands of dollars. I recommend that space always be centrally managed.

## Open versus Closed Floor Plans

Open floor plans are all the rage. These consist of large open rooms with numerous cubicles. There is usually a conference and team room for use by cubicle employees and offices for the managers. This method is said to save space while creating an open working environment that creates synergy and communication. Unfortunately, this is not the way it works out. These open floor plans are extremely noisy. Too much talking, radio playing, and general noise detracts from the environment. These employees usually make numerous complaints. These floor plans require the implementation of procedures to reduce noise. Special finishes can be purchased that absorb noise, and a noise generator can be installed that plays what is called white noise that blocks out unwanted noise. When this floor plan is used, it is supposed to have the main offices in the center core to allow light to enter the room. Unfortunately, that does not often happen. The senior employees usually put their offices on the windows, so they get the sunlight with very little light coming into the open spaces. In some facilities, the inner walls of the offices are made of glass so the light can be brought into the open space. Unfortunately again, the private office occupants want their privacy and usually frost the glass that dilutes the light coming into the open space.

A closed floor plan is the traditional use of offices and conference rooms. Office space is divided into offices utilized by the employees. The size of the mandate is determined by the rank of the person using it. This method requires the use of inner hallways to allow movement through the bureau. The inner hallways as you can imagine are wasted space. The inner offices have no windows. This method wastes space that costs thousands of dollars. It does not capitalize on the use of sunlight for the inner space.

There is no perfect solution. The method used should be determined by the needs of the organization. However, it is important to understand the pros and cons. Do not fall for the hype of the open floor plan to solve all your problems as well as saving resources. Space Planners must be prepared to explain the method to the unit and office manager and help them with the decision.

## Teleworking and Hoteling

There is much support for using telework and hoteling to improve employees' work life balance as well as save money by limiting commuting and facility costs. Telework is allowing employees to work from home using a computer connected to the net along with the telephone. This eliminates the need for this employee to travel to work each day. With the increased use of Wi-Fi, this employee could work anywhere with a Wi-Fi connection. Traditionally this employee will telework one or two days a week. Telework has its drawbacks. Use of sensitive work over the open net is prohibited because the possibility of a hack and potential loss of sensitive information. There are no rent savings unless space this employee used is given to someone else or used for another purpose. In addition, the employees that are teleworking are not available for walk-in customers at the office. These employees are also "out of sight, out of mind"." Many teleworkers believe they have been overlooked for promotions and other opportunities because they were not in the workplace every day. Telework can be a good thing or a bad thing depending on how it is used in the workplace, however, it seldom saves money in rent because people do not want to give up their physical space at work. This negates one of the primary reasons for implementing telework.

Hoteling is very similar to telework and is usually used at the same time. Hoteling uses whatever office use method the organization has in place. However, employees participating in the hoteling program do not have a specific workspace they use every day. Employees give up their workspace and must reserve space whenever they are in the building. Their unit or office pays rent on the actual time the workspace is used. The employee normally will not get the same space each time. They cannot leave belongings in the workspace. In many cases, employees come to work and use their old space without reserving it. This method requires the purchase and use of an automated software that takes reservations and charges the appropriate office. The Facility Management Office must also monitor space usage and ensure it is not used without reservation. This method is more trouble than it is worth. The two times I have been involved in this method, it was discontinued after about a year. I do not recommend using this application.

## Using Office Space for Furniture Storage

Many unit and office managers will want to use vacant office space to store extra furniture they may have. This is a waste of office space and should never be tolerated. Office space is just that. If the Facility Managers want to provide storage space for the unit and office managers, using the basement is a much better option. This is lower quality space with no windows where the Facility Manager can build out cages and rent them to the unit and office managers.

**Politics of Space**

There is a political aspect to space. In spite of some experts saying that people want collaborative space, I have not seen that. People feel they have earned a certain amount of space, a corner office, or a window. I still see people who want a 300 Sq. Ft. office with a conference room even though they could get their work done with less. The Space Manager cannot give everyone what they want. She or he can give everyone what they need. This is where a Sq. Ft. average helps. They can allow for an ego to be soothed by the window office while not letting the entire organization get more space than they need. Remember this when speaking with customers.

**Summary**

The proper management of space is important so that people are given the amount of space they need to work properly while paying the correct amount of rent. The cost to obtain and operate the facility is only second to the cost of human resources (Cotts, 2010). This makes the cost a critical element of business operations. In this chapter, my intent was to provide familiarization with space measurement, allocation, and provide other essential information. This was to help explain how space is managed effectively. Not understanding the basics can cost thousands of dollars in overpayments. The next chapter will explain how the space planner determines how much space each unit within the organization should get.

# Chapter 2 – Programming, Planning, and Budgeting

## Introduction

The Facility Manager operates as a Business Manager. "Better facility business planning and the reduction of churn offer the two best opportunities for facility reduction" (Cotts, 2010). A Facility Manager will likely be hired by an organization that already has facilities, and you will simply alter them to fit worker needs and maintain equipment so that all systems work as required. From time to time, you will be called upon to alter the facility to meet changes in worker needs. Your organization will depend on you to determine the actual requirements or what is required for space. You will usually assign this work to a Space Planner. Mark Karlen (1993) provides us with the basic steps of the process that include:

- Interviews.
- Observation of existing or similar facilities.
- Establish architectural parameters.
- Organize collect data, the first phase of program.
- Research unknowns.
- Analyze the data.
- Interpret and diagram the data, the complete program.
- Summarize the data into a finished document.

## Requirements

There should be an old program of requirements that explains the amount and type of space the organization needs. This should be updated every three years. If there is not one, you should contract for one to be done.

Requirements start by identifying the actual number of workers. Second is to identify actual space that the company will provide for each worker. Each company develops its guidance, and this is not one of those topics where there is a right or wrong answer. The decision is greatly affected by the amount of money that is available for rent, utilities, furniture, fixtures, and equipment. In general executives get more space than the manager. Managers get more space than workers. There is also space needed to walk around the cubicles between offices that is called circulation. I like to use

35% for circulation space. Some prefer to use 25 or 30%. To calculate the total amount of space, you add the percentage of circulation space to the actual space. It is important to take a minute to address windows. Normally windows in commercial office buildings do not open. Their purpose is to allow sunlight to come into the facility. Usually, windows are given to executives and senior managers and are considered perks by those that get them. In general, I like to start with the guidelines in Table 2. It is common to work with averages to begin with. I like to use a 200 usable Sq. Ft. average. This allows the executive to have a bigger office, space for a waiting room and conference room, offices for higher paid workers, while others work in cubicles. An example could be for two executives, a director, and her deputy. Each gets 300 Sq. Ft. office, a 100 Sq. Ft. waiting area, and a 225 Sq. Ft. conference room. There are also two human resource managers that get a 200 Sq. Ft. office and fourteen human resources specialists. Six of the human resources specialists need a private office because of privacy issues. The other eight get a 64 Sq. Ft. cubicle. There are three contractor IT specialists that get 42 Sq. Ft. cubicles.

| Position | Square Footage (Sq. Ft.) |
| --- | --- |
| Executives | 300 |
| Managers | 225 |
| Supervisors | 200 |
| Analysts | 150 |
| Human Resources Specialists | 150 |
| Budget Staff | 150 |
| General Workers | 100 (office) or   64 (cubicle) |
| Contractors | 42 (cubicle) |
| Conference Room | 225 |
| Waiting Area | 100 |
| Storage and file rooms | 100 |

Table 2 – Office Size Guidelines

The Human Resources group also needs a 100 Sq. Ft. storage room. The total square footage for this office is 4,000 Sq. Ft. This is outlined in Table 3. After you know this information, you need to find out if there is anything special that is required of any spaces. This could be such things as windows, public entrance, large file systems, additional security, or installation of IT equipment.

Best practice is to build these tables in Microsoft Excel® as spreadsheets. This allows the use of formulas in the cells so that the spreadsheet does the real work. Some use a pivot table so that the basic data is updated as needed, which populates the table with updated data.

| Human Resources Office Requirements | |
|---|---|
| **Position** | **Square Footage (Sq. Ft.)** |
| Director | 300 |
| Deputy Director | 300 |
| Waiting Room | 100 |
| Storage and file rooms | 100 |
| Manager 1 | 200 |
| Manager 2 | 200 |
| Human Resources Specialists with Privacy Need (6 persons) | 900 (offices) |
| Human Resources Specialist with no Privacy Need (8 persons) | 512 (cubicles) |
| Contractor IT Specialists | 126 (cubicles) |
| Conference Room | 225 |
| Subtotal | 2,963 |
| Circulation Space | 1,037 |
| Total | 4,000 |

**Table 3 – Example Human Resources Office**

## Adjacencies

The next step is to do an organizational bubble diagram that indicates the relative importance of the relationships between individuals, groups, and departments to help in deciding where to locate them (Addi and Lytle, 2000). The International Facility Management Association defines this as an adjacency diagram that conveys the desired proximity of workspace elements or functions to each other (Facility Management Glossary, 2013). I have given you both terms because both are often used, but I prefer adjacency diagrams. You begin with an organizational chart. You draw a circle for every part of the organization. The size of the circle is in direct correlation to the Sq. Ft. of the room.

That is step one of the adjacency diagram. Step two is to draw lines connecting the groups of people that need to be adjacent to each other. The required adjacencies are in a bold line between the two circles. The nice-to have, yet optional, adjacencies are in a thinner line between the two circles. For example, a diagram for Table 3 will look like Figure 1 on the next page.

**Figure 1 – Adjacency Diagram**

Figure 2 shows how the diagram would look with the lines connecting offices that need to speak and work together frequently.

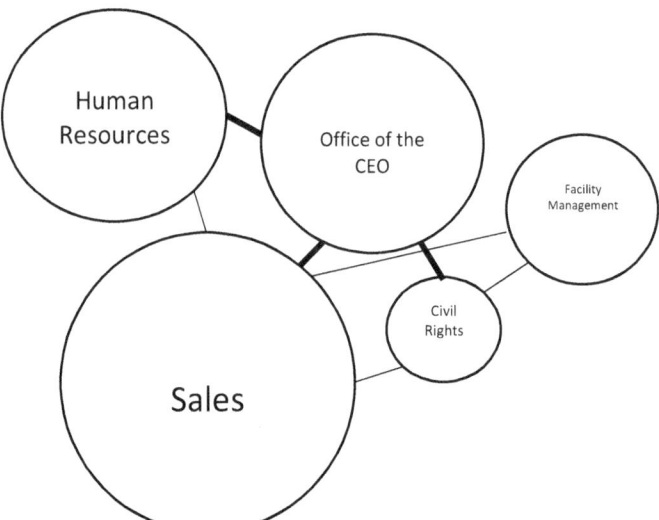

**Figure 2 – Adjacency Diagram with importance of adjacency lines**

Now an adjacency diagram is done for the Human Resources Office. The diagram would look like Figure 3.

**Figure 3 – Adjacency Diagram for Human Resources Office**

The adjacency diagram for the Human Resources Office with the important adjacency lines would look like Figure 4.

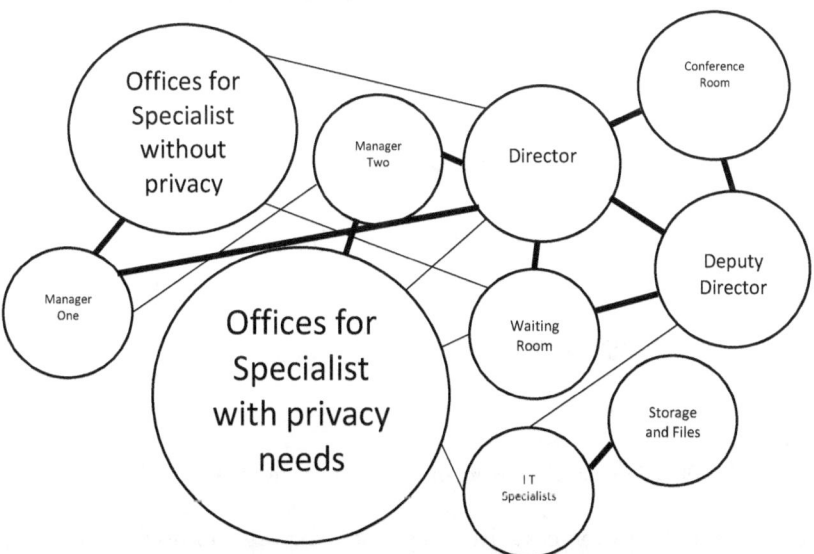

**Figure 4 – Adjacency Diagram for Human Resources Office with importance of adjacency lines**

**Blocking and Stacking**

From this, we could layout a floor plan to put these people together on the same floor next to each other. That process is called blocking and stacking. Blocking is putting a group of people together on the same floor. A Facility Manager should be concerned with stacking plans or how unit and offices relate to each other in a multi-floor building (Cotts, 2010). Stacking is putting a group of people together on the floor above and below each other. Either way is acceptable. One way of doing this is to mark up a floor plan with pen and ink changes that would be put into an automated drafting program at a later date.

In Figure 5 we have a floor that has enough Sq. Ft. to cover the entire organization we used in Figures 1 and 2. In this case, we would lay out the units within the organization so that the most important adjacencies are met. In this instance, we can accommodate all adjacencies identified in the analysis. We use a cloud to mark the space that would be utilized by each unit of the organization.

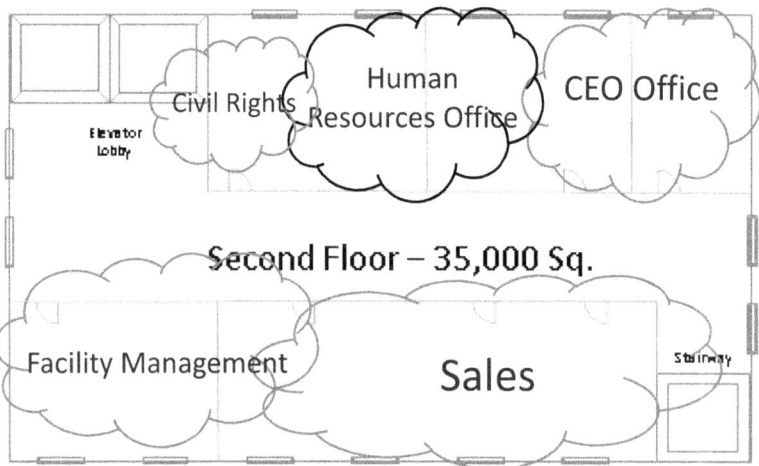

Figure 5 – Blocking plan for the organization on one floor.

This is not always the case. In many cases, the organization would be split over floors. This is a more difficult process. In a stacking plan, we would look at two or more floors to get us the amount of square footage we need. In commercial buildings, the floors are usually very close to the same square footage. So we would look at the number of floors to meet our needs. In a building that has two floors that meet the requirements, we would put the offices by the adjacency requirements. Mark Karlen (1993) describes bubble diagrams and block plans as the first planning steps.

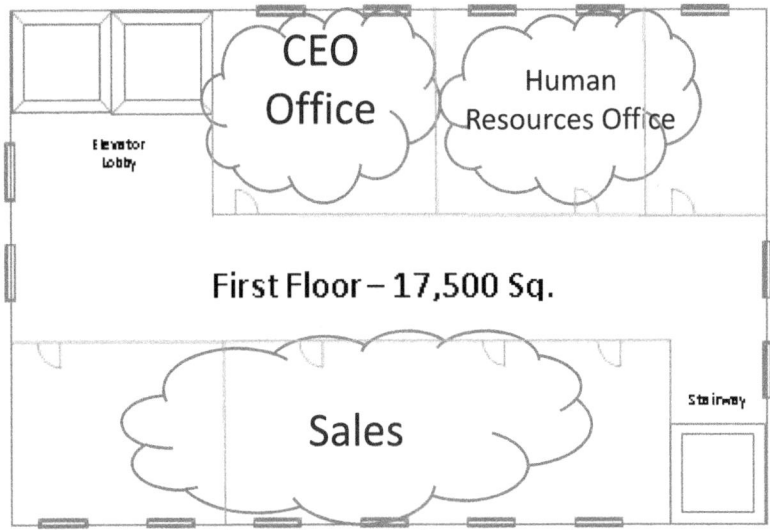

Figure 6 – Stacking plan for the organization using two floors.

I have found that blocking plans also show the customer where space would be located on the current floor plans. Initially using a tracing paper over the floor plan can save time and effort. The tracing paper allows several options to be drawn.

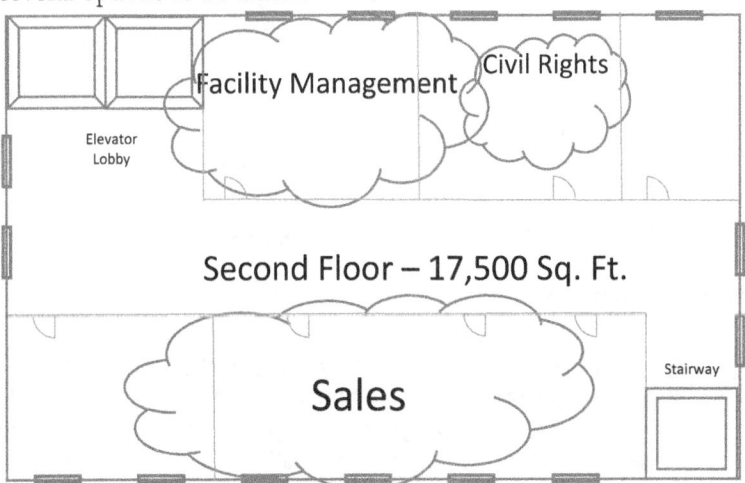

Figure 7 – Stacking plan for the organization using two floors.

Once we develop the plan for the organization we go down a level to the unit. If we look at the Human Resources Office, we will layout the unit space similar to that shown in Figure 8.

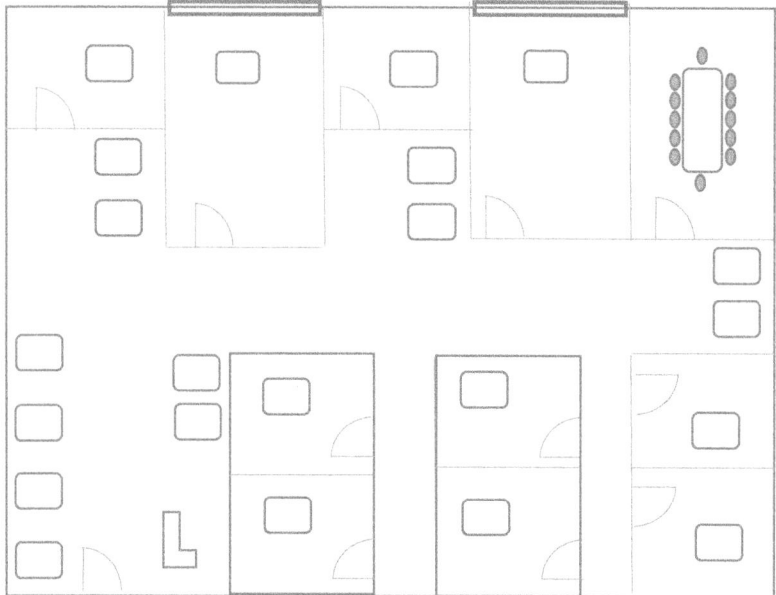

**Figure 8 – Human Resources Office layout.**

The process continues with each unit's space being identified. This is a long and arduous process that can take weeks to complete, depending on the size of the organization.

Once all space has been identified, and each unit is placed into space the total cost is determined. The actual costs are defined and broken down by into some Sq. Ft. each unit occupies. The cost of maintenance is also identified by accessing industry standards for the type of facility.

## Requirements Document

I want to raise a couple of notes here to the square footage determined to be needed by the organization. Swing space, or space needed while alterations and renovations are done, should be estimated at 2-3% more than the estimated square footage required annually for large organizations and 5-7% for smaller organizations. It is also important to add 10% to any amount of space that has been identified to account for minor changes in space needs. With the need and cost identified and preliminary plans developed, the designer now steps in to put all the programming and planning into the as-built drawings of the building to represent a set of construction drawings that can be used to build the space for use. A complete set of Construction Documents, or CDs, will consist of the following drawings:

- Architectural – Floor Plan, Elevation, Section, and Details.
- Electrical – Plan, Elevation, and Details.
- Telecommunications – Voice, Data, and LAN.
- Cables – TV, Audio-Visual.
- Wall Finishes – Paint, Wall Covering, Fabric, Wrapped Panel or Sound Soak. All finishes are standardized by building.
- Flooring – Carpet (broadband or carpet tile), VCT or Access Flooring. All finishes are standardized by building.
- Reflected Ceiling Plans.
- Mechanical – relocate sprinkler head(s), supplemental A/C.
- Specifications (including Technical Data).
- Independent Cost Estimate (ICE).

The Space Planner must ensure that the CDs comply with:

- Fire Protection.
- Life Safety.
- Physical and technical security procedures and requirements.
- Americans with Disabilities Act (ADA) and Uniform Federal Accessibility Standard (UFAS) requirements.
- Mandated space utilization.
- Other requirements.

Proceed with construction documents consisting of design, specifications, and ICE. Add ICE cost to log book.

Submit one (1) copy of CDs to the Space Manager for review, comment, and approval. Revise the drawings as required. All of this information is put into a single document with floor plans and all that is in the Requirements Document for the organization. This is usually printed in 11" x 17" size and is maintained on file.

## Budgeting

It is essential that budgets are prepared for future years to ensure that facility operations are sustainable. It is common to go three and five years into the future. The current year budget is said to be acted upon. Meaning it is being spent at that time. One year into the future is being adjusted so that it is more accurate and years three through five are being made based on estimates of increases in expenses, see Table 4 for example. The most common categories of the facility budget include:

- Payroll.
- Rent or credit payments.
- Taxes.
- Utilities.
- Logistics support contract.
- Support contracts.
- Telephone services.
- Capital projects.

In rented facilities the cost of rent will usually go up 3-5% annually. The taxes and maintenance cost shared by tenants will normally go up as well. If the lease is up the cost increase for a follow-on lease can be 7-10%. This is what is considered when developing the budget in future years. If the facility is owned, it was likely financed. The payments for the credit are likely stable over the coming years. However, taxes and maintenance costs will increase.

Any and all contracts have an annual increase built into the future years of a contract. For example, a contract is usually for five years. Year one is the year the contract is enacted. Within the contract will be specific cost increases for year two, three, four and five. The budget for the cost is outlined in the year of the contract as they go into the future.

Utilities can go up anywhere from 3-7%. The usage should be known from past years. Future budgets can take the average usage and apply the percentage that is expected to the cost to determine the annual cost of utilities.

Telephone costs include local area networks, equipment rental, and lines run through the facility as well as minutes used. The usage charges are unique to usage in any given month, so this amount is hard to budget for. It is best to budget what the cost is likely to be.

Capital projects are determined by the organization and should be spread out over the five-year period. Capital projects also result from business cases. Capital projects are usually used to complete replacement of mechanical, electrical, or plumbing systems. This replacement is due to the age of equipment and is planned years ahead of time. Environmental damage from a tornado may also come to the level of capital projects and would not be budgeted for. Capital projects have annual cost implications with respect to depreciation and operations (Cotts, 2010).

| Item | 2015 | 2016 | 2017 | 2018 | 2019 |
|---|---|---|---|---|---|
| Facilities Payroll | $829,012 | $845,592 | $870,960 | $888,379 | $915,031 |
| Rent or Credit Payments | $1,289,000 | $1,353,450 | $1,421,123 | $1,492,179 | $1,566,789 |
| Taxes | $35,256 | $35,256 | $35,256 | $37,678 | $37,678 |
| Utilities | $65,453 | $68,726 | $72,162 | $75,770 | $79,559 |
| Logistics Support Contract | $895,345 | $958,019 | $1,025,080 | $1,096,836 | $1,173,615 |
| Support Contracts | $91,245 | $97,632 | $104,466 | $111,779 | $119,603 |
| Telephone Services | $63,465 | $67,908 | $72,661 | $77,747 | $83,190 |
| Capital Projects | $185,000 | $95,000 | $105,124 | $86,000 | $75,000 |
| Total Facility Budget | $3,453,776 | $3,460,473 | $3,709,832 | $3,866,368 | $4,050,465 |

**Table 4 – Example Facility Management Office Budget**

The Facility Manager should plan 5-15% more for discretionary work than budgeted (Cotts, 2010). The budget provides a plan for the work to be accomplished by the Facility Management Office annually. Work that is not funded cannot be done. If work is not funded for more than one year damage can begin to take place. If this situation persists equipment life can be greatly shortened, or it can fail at the most inopportune time. Either situation usually costs more money than if the equipment preventive maintenance and repair had been done on time.

**Chargeback System**

It is best for the Facility Manager to charge for space and services. The charges are in two areas referred to as above and below standard services. The customer is charged a specific rent that comes with a standard level of services. Any services above these are called above standard level services. For above standard services, the Facility Manager should develop a financial method known as a chargeback System. When a unit or office requests above standard services they are provided a cost estimate. They approve that cost estimate, and after the work is done, they are charged the actual cost of the service. "If chargebacks are used expend the effort to make them meaningful. The concept of a fixed package of services for a relative internal rate (with services outside the package provided on a fee-for-service basis) is the best practice" (Cotts, 2010). Let me give you two examples.

- Within the standard level of services, the facility is cooled or heated from a set hour in the morning to a set hour in the early evening. If a unit or office want cooling or heating before or after those hours they pay for the actual amount of the cooling or heating.
- A second example is that a unit or office wants to alter its space to separate offices. The Space Planner develops requirements for the work and provides the unit or office manager with an estimate. The unit or office manager approves the estimate and after construction the actual costs are billed to the customer.

The chargeback is done monthly or quarterly. Chargeback is gathered throughout the period and submitted to the budget office. The budget office transfers the amount of money from the customer to the Facility Management Office.

**Summary**

It is important to remember that the configuration of the building influences space planning (Karlen, 1993). Most buildings have columns, elevators, and vertical space that must be planned around. The work described in this chapter must be reviewed and validated annually. This review will ensure that space is being properly used, and costs are within budget. Changes are made based on the findings of the review. These documents are crucial to the organization and must be maintained to ensure space is properly used. Aside from emergencies, the work that is carried out by the Facility Management Office is based on these documents.

# Chapter 3 – Drawings and Floor Plans

## Introduction

The Facility Management Office keeps floor plans to show the current build-out of space. These drawings also show where equipment is located as well as where services like gas, electricity, and water were placed. These floor plans are used by maintenance and alterations personnel alike. To be useful, they must be correct, which takes time and effort from everyone involved. These floor plans are generated and maintained in an automated drafting software program on a computer. The drawing and floor plans are done by a draftsperson or designer and must meet several codes to be correct. These occupations require years of study and are considered experts in their field.

## As-Built Drawings

The basic set of floor plans that every Facility Management Office should have is a set of the original drawings that were used to build the facility. These are referred to as As-Built Drawings. A member of the Facility Management Office is usually tasked with updating the As-Built Drawings as the facility is changed. Before any alterations, a Space Planner must verify field conditions. Any changes identified during this site visit should be annotated on the drawing. The Space Planner noting the changes must make those changes on the As-Built Drawings to keep them current. These drawings are used by the maintenance side of Facility Management to know where equipment and assets are located. The drawings will have the entire mechanical, electrical and plumbing system included in the drawings. Whenever equipment needs to be replaced or emergency shut offs need to be located, these drawings are consulted. Failing to keep them up-to-date can mean damage and loss of service to facility occupants. These drawings are enormous and are usually maintained in a computer automated drafting software. Printing them out can be costly and is very seldom done. It is more prevalent to print out a particular page or two when needed. The Facility Management Office has desktop computers equipped with the computer automated drawing software. They also have a plotter that is capable of printing large format drawings.

## Blocking and Stacking Plans

The second set of drawings that must be developed and maintained are the Blocking and Stacking Plans. This is a set of drawings that were developed from the requirements identified for the space. They do not include mechanical, electrical, and plumbing. The purpose of these floor plans is to identify all offices listed in the print. The plans identify which units within the organization are located in which offices. It is important for units to be contiguous and to be adjacent to other units they work with. A specific color normally identifies a unit, see examples at Figures 9 and 10.

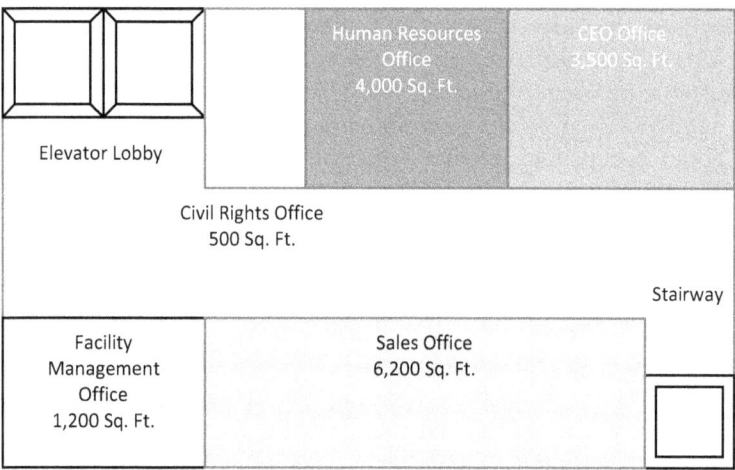

**Figure 9 – Example Blocking Plan**

Once all units are shown on the drawing it is obvious where each unit is and whether they are contiguous and adjacent. As you can imagine blocking and stacking plans are a snapshot in time. Almost as soon as they are printed, offices and people move and they are no longer correct. In spite of that drawback, it is essential to do Blocking and Stacking Plans every five years and to try to move offices and people back to being contiguous. These plans are most important when considering moving to another facility. These plans show how much space is needed and allows the Facility Management Office to rent just enough space at the right price. This prevents renting of unneeded space, which wastes money.

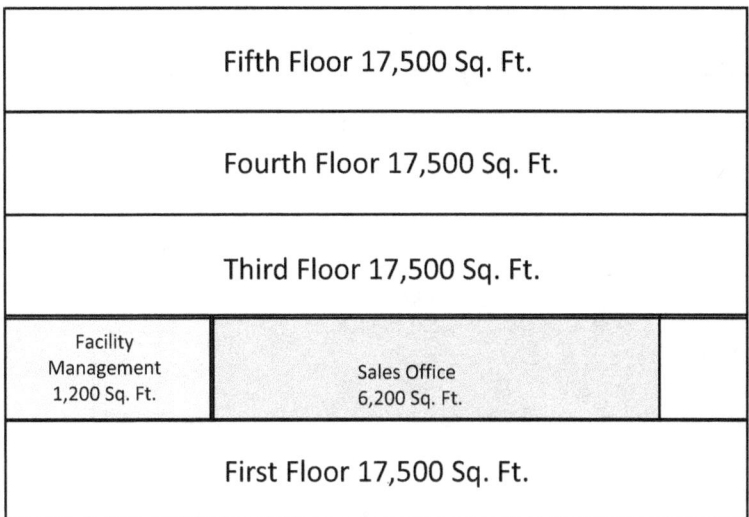

**Figure 10 – Example Stacking Plan**

From this floor, plan changes can be determined, and plans can be made to move offices to correct any issues. As I said earlier, an organization is usually properly blocked and stacked when it first occupies a facility. Once you move units and workers the blocking and stacking plan changes and create issues. It is usually too expensive and time-consuming to correct issues with blocking and stacking any more often than every five years.

**Key Plans**

The third set of drawings that must be developed and maintained are Key Plans. Key Plans are miniature plans of facilities scaled to fit on normal sized papers. I like to use 8½" x 11" or 8½" x 14" paper. However, 11" x 17" is often used. Any of these can be printed on the average office printer. These key plans must be developed and kept up-to-date to know where units are located within a facility. Throughout the year, Space Planners maintain a database to track space allocations and assignments by building, room in building, and by a unit of the organization. The Key Plans are first developed in full-size computer automated drawing software but are saved as smaller Adobe PDF files to make them easier to use, see example at Figure 11.

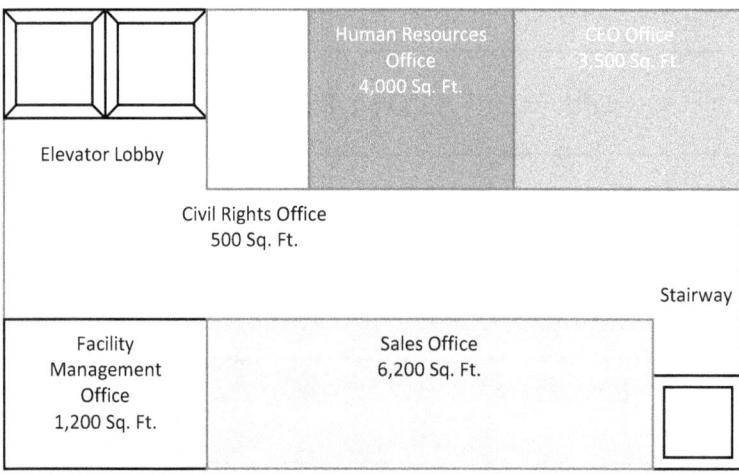

**Figure 11 – Key Plan Example.**

These plans are done initially and at some frequency determined by the Facility Manager. I have done them quarterly, semiannually, and annually. I think they work fine when done annually. The Space Planner conducts an annual site inspection to verify space allocations and assignments. The Key Plans are revised to reflect the changes.

**Fit Tests**

There is a particular kind of drawing that is done to ensure that people and furniture will fit into space. These drawings are called fit tests. These are used when an organization plans to move from one location in a facility to another. The new space is drawn to scale with the actual furniture. These drawings will tell the unit or office manager if the furniture and people will fit into the new space. These drawings save the time and frustration of moving into new space only to find out that the furniture and people do not fit. These drawings are an essential part of any move.

**Summary**

It is critical that the various types of floor plans be kept and maintained. It is from these plans that maintenance and alterations staff maintain the facility in good repair and build it out to meet the customer's needs. However, to be useful they must be correct, which takes time and effort from everyone involved.

# Chapter 4 – Owning versus Leasing

## Introduction

As a Facility Manager, your organization will have the choice to own or lease space to house workers, customers, files, and equipment. There are pros and cons to each method that depend on the company's specific needs and resources. Before making a choice it is important to identify the actual requirements or what is needed.

In Chapters 2 and 3 I explained how to determine the utilization rate of offices and how to develop the requirements needed for facilities. Through this process, the amount of space needed by an organization is determined. In this chapter I would like to touch on the hypothesis that forecasting space needs can be done with two basic strategies (Cotts, 2010):

1. Occupy owned space, which permits maximum control.
2. Occupy leased space, which permits maximum flexibility.

The result of this effort is that organization leaders need to make what is called a lease or buy decision. The majority of corporations choose to own rather than lease facilities (Cotts, 2010).

## Leasing

If the decision is made to lease a facility, the Facility Manager will be asked to obtain space for lease. I have not worked in an office that had the expertise to obtain leased space. Therefore, I recommend contracting with a commercial real estate firm to do all the work to lease a facility for your organization. The Facility Manager approves and signs the lease. A commercial lease can be very expensive. It can cost millions of dollars over the life of the lease.

You might lease an entire building or part of a building. Basic commercial real estate principles for your state apply to all leases, but there is some general information I can provide here to help familiarize you with the leasing process. First there are different types of leases. I outline them in the following paragraphs. The basis for rent charges is the rentable Sq. Ft.

Do not let anyone convince you to pay for gross Sq. Ft. You should only pay for the rentable amount of space. The various types of leases include:

The gross lease is similar to a full-service lease. The difference is not significant, and most people consider them together. The landlord pays for taxes, insurance, and maintenance. The gross commercial lease is used most often in multi-tenant and single tenant office buildings, industrial and some retail properties. The landlord collects fixed rents and pays the expenses (The Gross, 2015). Since costs increase over time, many gross and full-service leases contain escalation clauses that increase rents over time to offset tax increases and higher insurance and maintenance costs (The Gross, 2015). It is important that you understand any escalation clauses to project rent expenses into the future.

A net lease is a type of commercial real estate lease in which the tenant pays for their space, as well as for part of all "usual costs" that the landlord pays. Expenses incorporated into net leases may include taxes, utilities, janitorial services, property insurance, property management fees, sewer, water, and trash collection. Net leases almost always favor the landlord and should be negotiated to include caps, or, the maximum amount a landlord can increase fees each year (What, 2015).

- A double net lease is a type of net lease in which the tenant pays all or part of taxes and insurance associated with the use of the property. These fees are paid in addition to monthly rent for use of the actual space.
- The triple net lease requires you to pay a significant share of the expenses of operating the facility that include taxes and insurance. This type of lease helps the landlord by fixing their costs, as their rents are fixed (Types, 2015). Tenants are not fond of this kind of lease, especially in older properties. The triple net lease is used extensively in commercial real estate. It is popular for multi-tenant industrial and retail properties (3 Most, 2015). With tenants whose expenses vary widely, such as an industrial user of electricity, the triple net lease is best for the landlord. Tenants are resistant to triple net leases because they have no control over increases in expenses and budgeting their costs is more difficult. This is especially true when it comes to repairs and maintenance. In a triple net lease, the tenants would be responsible for sharing the cost of roof replacement. This can be a large and unexpected expense. Of course, fixed rent is lower with the triple net lease. If the building is a newer one, tenants may find the triple net to be preferable to other choices. If establishing a new business, the triple

net tenant in a new building can enjoy lower rent and expenses in their first few years (3 Most, 2015).

- The modified net lease is a compromise between the gross lease and the triple net lease. This lease helps landlords and tenants structure lease terms that work for both. The landlord and tenant usually set up a split of maintenance expenses, while the tenant agrees to pay taxes and insurance. Utilities are also likely negotiated in the modified net lease. This type of lease might be used in industrial, retail or multi-tenant office properties. Tenants should resist the triple net leases in older properties.015).

The length of the lease depends on the plans for your organization. The longer the lease, the cheaper the actual rent per Sq. Ft. is. That is because the landlord gets a stable income over an extended period. A short lease does not provide this stability. Therefore, the landlord charges more. One method of leasing is to agree to a base set of years with options. For example, you could have a base lease of five years with two five-year options for a total of 15 years. The first five years are all that is agreed to up front. In the fourth year of the first five years, you decide whether or not to pick up the option for the second five years. At this point, you can elect not to pick up the option for the second five years. There is no penalty for not picking up an option. The procedure is the same for the third option.

The lease will also include building or adapting the space to meet your office needs. This expense is usually amortized in the rent or can be paid up front. You agree in the lease that the landlord will design and build out space for a specific amount of money per Sq. Ft. The build out can cost on average $250 per Sq. Ft. This can also run into the millions of dollars. You also need to get agreement on returning space to its original condition. I do not recommend you ever pay to return space to its original condition. Within the lease will be timelines that you will have to meet for reviewing and approving designs and accepting space when completed. Complete is usually referred to as "substantially complete:" This means that the workspacc is complete enough to allow workers to move in and work from the space. This does not mean it is complete. There may be some minor work going on that is referred to as correcting "punch list" items. You will also have to coordinate the phone and local area network installation as part of this construction. Furthermore, you will need to coordinate furniture installation and personnel moves to get people in the building efficiently. You will start paying rent when space is "substantially complete."

The lease is a legal document and must be reviewed by a lawyer before approving to ensure that what you want is what is in the lease.

## Owning

If the decision is made to purchase a facility, the Facility Manager will be asked to obtain space for purchase. I have not worked in an office that had the expertise to buy space. Therefore, I recommend contracting with a commercial real estate firm to do all the work to obtain land for the facility and an architect and engineering firm to design and build the facility. The Facility Manager approves and signs all documents. Purchasing commercial buildings or even part of a commercial building is very expensive. The purchase usually runs in the millions of dollars. Once land is purchased the building will have to be designed and built to the needs of the organization. The build out can cost on average $250 per Sq. Ft. to build out. This can also run into the millions of dollars.

Building a commercial building can take up to a year from beginning to end. The process starts with requirements developed for the organization. The architect and engineering firm will take those requirements and determine the size of facility needed. The real estate irm takes this information, and a location is found for the new facility. Determining the location can be very time-consuming. The particular location is dependent on the needs of your organization and the amount of money that is available for the facility. The location must also be zoned for your facility use. Some places may be cost prohibitive due to utility and easement expenses that must be paid. This includes bringing water, sewer, electricity, telephone lines, and roads to the facility.

While the location is being selected the architect and engineering firm will be designing the facility. The design is based on the requirements document and the architect's personal design. The main focus of this step is to ensure the design does not exceed the budgeted amount.

Designing the building to meet the requirements of the organization is a must. From that construction documents are developed, and contracts are issued for the actual construction. Permits must be obtained to allow for the building to be built. These are obtained from the local municipalities and can take months to get. The first set of drawings the Facility Manager will review is at the 35% phase. This includes the basic layout of the building, with mechanical, electrical, and plumbing. After this, a set of 65% drawings is completed that makes corrections identified in the 35%.

Finishes will usually be included in the 65%. The next set of drawings is the 100%. This set of drawings includes corrections identified in the 65% drawings and is a complete set of drawings that includes a cost estimate and is used to contract and build the facility.

When purchasing or building a facility the organization usually gets credit to pay most of the cost. The money is paid back each month at a predetermined interest rate plus the principal. The payback period is imperative and must be agreed to by the financial officer of the organization. Obtaining financing can be very tricky and must be done at an interest rate that the organization can afford.

Part of the design is the interior build out that will include furniture, finishing, and equipment. Furniture costs can be very high. Style and use are considered along with the utilization rate. Normal office furniture is contemporary style and consists of a desk, bookcase, credenza, executive chair, and two visitor chairs. Standard office furniture can cost $5,000.00 per office. Executives get executive style furniture that consists of a desk, two bookcases, credenza, executive chair, round table with four executive chairs. Executive office furniture can cost $8,000.00 per office.

General workers usually get cubicles or what is commonly known as systems furniture. This is furniture that is made from standard pieces in a variety of configurations to suite the customer's needs. Standard cubicles can range from 6 Ft. x 6 Ft. to 8 Ft. x 10 Ft. The more popular sizes are 8 Ft. x 8 Ft. that give a worker 64 Sq. Ft. of space. A cubicle usually consists of a work surface, overhead storage cabinets, electrical outlets, one rolling file, executive chair, and three walls. This setup usually runs about $6,000.00. I like to include a coat locker that provides space for the worker to place their personal items. I like to have the rolling file with a cushion on it so it can be used as a chair. The rolling file usually has one file drawer with two accessory drawers. Doors can be purchased for cubicles if needed. The height of the typical cubicle is 48 inches. I think this is a good height, but some work situations require a higher wall. In this case, I usually make the height above 48 inches with a frosted glass panel. This gives the additional height without blocking out the light. These additional items increase the cost of the systems furniture. Systems furniture is also designed to fit into space. This design is done once a 35% design of the facility is finished. This design includes the electrical, telephone, and local area network within the systems furniture. Owning is cheaper over the long-term (Cotts, 2010).

## Summary

The Facility Manager is responsible for obtaining facilities to meet organizational needs. The two most likely methods to get a facility are to lease one or buy one. There are pros and cons to each method. Identifying the actual requirements is the beginning of either method. The result of this effort is that organization leaders need to make what is called a lease or buy decision. Experts "urge the facility management department to work closely with finance, counsel, and top management in the lease versus purchase decision" (Cotts, 2010).

# Chapter 5 – Altering Space

## Introduction

Days after the organization moves into a new building, or a newly leased building, things begin to change. People or offices are moved to different locations that they think are better. This is called churn or the rate of people moving within the organization over time. In offices I worked in, we moved hundreds of people every week. This is what is called a high churn. You cannot avoid churn, but you should do as much as you can to control it.

## Process

Control begins with a formal requisition system for space planning services that includes basic alterations. In larger Facility Management Offices, alterations are handled by a Space Planner, who will design and draw changes. The work is then picked up by a Facility Specialist, who will contract for the work and see it through to completion. The Facility Specialist will confirm the work is complete and accepted by the customer and will pay all contractors and charge the client for the work. In a smaller Facility Management Office, this work will be done by one person who acts as the Space Planner and Facility Specialist. The Space Planner is usually adept at design and uses a computer automated design software to complete drawings. They are also familiar with color and finishes. The Facility Specialist is adept at contracting, construction management, and financial processes of billing and paying for work completed.

A customer can be anyone in the organization who can authorize charges against her office or unit. When the client wants or needs alterations or renovations, they submit a request that will be reviewed by a Space Planner in the Facility Management Office.

## Concepts and Designs

The Space Planner will print out the existing office area and meet with the customer to determine the actual changes that are being requested. Pen and ink changes are made to the existing floor plan. From this, drawings are done that show how the space will look after it is renovated. The Space Manager centrally manages drawing numbers. The drawing numbers are kept in a computer file for easy access. For each drawing number, the Space Planner fills out a log with information such as the requesting organization's name, room numbers, and Sq. Ft., type of space, and the date the drawing number is assigned. The Space Planner will then ask the Space Manager for approval to proceed. Once approved the Space Planner proceeds to change the drawings and returns to the drawing number log to note the date the drawings are completed. Not all alterations and renovations are the same size and complexity. The harder the job, the longer it normally takes to complete and the more it costs.

The Space Planner verifies the existing conditions and updates the as-built drawings as necessary. This means the Space Planner must go to the location of the proposed work and compare the existing conditions with the as-built drawings. If conditions agree, the Space Planner can proceed with making changes. If the drawings and existing conditions do not agree, the Space Planner must make pen and ink changes on the as-built drawings and make the changes to the drawings in the automated design software.

Once existing conditions have been verified, the Space Planner meets again with the customer. This second meeting is used to verify the client's request and determine additional requirements that may be necessary. The Space Planner develops conceptual designs based on client requirements. If necessary, the Space Planner provides the client with alternatives or recommendations. If the client requirements are unreasonable or clearly depict waste and abuse, e.g., demolish five offices with the largest office in the corner to build five offices that relocates the largest office to the center, the Space Planner must get the Space Manager's approval.

The Space Planner checks with the client to determine if space will be vacant during the construction. If not is swing space available during the construction. Swing space is a location that workers are moved to while construction is being done. If the office or unit does not have swing space, the Space Manager will try to locate space for this use somewhere else in the facility. The Space Planner also determines if the project is planned for one or several phases. Once the concept is developed, it is put into floor plans and is submitted to the client for review, comment, and approval. The Space Planner then incorporates changes requested by the client into the drawings.

If the client needs to procure or acquire new furniture, the Space Planner provides furniture configurations; however, if systems furniture is required, the Space Planner prepares a schematic or block diagram of the furniture layout. A "best practice" for systems furniture is to limit selection to three types and three different types of finishes. This allows the furniture to be repaired and maintained. Also, the Space Planner coordinates with a Procurement Specialist to contract for the design, purchase, and installation of systems furniture. Space Planner review and comment on the furniture drawings and specifications prepared by the furniture designer. When the Space Planner approves the furniture design, it is sent to the customer for review and approval. The Space Planner coordinates with an engineer for all mechanical, electrical, and plumbing support. It is essential that the client pays for modifications to the mechanical, electrical, and plumbing systems made for their renovation. The Facility Management Office usually does not budget for these types of modifications.

For video conferencing support, the Space Planner contacts and coordinates with the video conferencing personnel for procurement and installation. For carpet support, the Space Planner contacts and coordinates with the Procurement Specialist. The Space Planner submits the carpet services requirements (i.e., color and quantity), the installation date, and cost estimate to the Procurement Specialist. A "best practice" on the carpet is to limit the types and colors of carpet to three to allow the replacement and repair to be sustainable. The Procurement Specialist contacts the installer to schedule the installation. For telephone, local area network, and data requirements, the Space Planner contacts, and coordinates with the telephone company. After completing the necessary contacts and coordination, the Space Planner provides all information to the customer who must approve final plans. Once customer approval is given, the full set of CDs is done. A complete set of CDs consists of the following drawings:

- Architectural – Floor Plan, Elevation, Section, and Details.
- Electrical – Plan, Elevation, and Details.
- Telecommunications – Voice, Data, and LAN.
- Cables – TV, Audio-Visual.
- Wall Finishes – Paint, Wall Covering, Fabric, Wrapped Panel or Sound Soak. All finishes are standardized by building.
- Flooring – Carpet (broadband or carpet tile), VCT or Access Flooring. All finishes are standardized by building.
- Reflected Ceiling Plans.
- Mechanical – relocate sprinkler head(s), supplemental A/C.
- Specifications (including Technical Data).
- Independent Cost Estimate (ICE).

The Space Planner must ensure that the CDs comply with:

- Fire Protection Code.
- Life Safety Code.
- Physical and technical security procedures and requirements.
- ADA and UFAS requirements.
- Mandated space utilization.

The Space Planner adds the ICE to the drawing log for that drawing number. The Space Planner submits two copies of the CDs to the customer for review and approval. The client should indicate "APPROVED" on one copy and sign, date, and return to the Space Planner. The Space Planner then submits three copies of the CDs to the Facility Manager and a copy to any personnel involved (e.g., carpet contractor and telephone company). The Facility Manager reviews the package and returns it to the Space Planner if they have questions or comments. If the Facility Manager approves, they will assign the project to a Facility Specialist and forward all CDs to them.

## Construction

The Facility Specialist works with the Procurement Specialist to contract for the work in the CDs. The Space Planner supports and assists the Facility Specialist with any clarifications and contractor requests for drawings. If the contractor's bids are over 30% higher than the ICE prepared by the Space Planner, the Facility Specialist should request that the Space Planner review the contractor's bid. The Space Planner should cooperate fully. The Procurement Specialist normally conducts a walkthrough with potential contractors and the Facility Specialist. The Space Planner is invited to the walkthrough to answer design-related questions raised by the contractors.

Once the project is contracted, the Facility Specialist oversees all construction, installations, and final inspections. Once the work begins, the Facility Specialist will conduct daily inspections of the construction site. They will notify the contractor of any discrepancies or activities being conducted outside the scope of the work and ensure that the contractor is following all guidelines for safety and health and that all areas are clean upon completion of work for that day.

The Facility Specialist uses a form to document the work order number, a brief description of the services and supplies provided by the contractor, the contractor's name, and the project cost. The Facility Specialist signs and dates the form to certify that the work requested is satisfactory and is accepted for payment. The Facility Specialist forwards this to the budget office, which will make the payment.

## Mechanical, Electrical and Plumbing Costs

In some organizations, unit and office managers refuse to pay for necessary upgrades and changes to the mechanical, electrical and plumbing systems caused by their alterations or renovations. This leads to long-term problems for the entire facility. Every alteration or renovation must include an analysis of the mechanical, electrical, and plumbing systems to determine the impact the alteration or renovation will have on them. If changes and upgrades are needed the unit or office manager must be told, and the cost included in the alterations or renovations. A couple of good examples include:

- The unit manager wants to take a large open floor plan and build out offices for each of his employees. The change from the open office plan to individual offices will directly affect the movement of air within the space. A mechanical assessment might indicate that a variable air volume box be added to move additional air to the offices. Without this, the offices will probably suffer from a lack of air movement. Complaints will likely result from hot and cold spots in the space as well as complaints from odors. The unit or office manager is apprised of the additional costs and refuses to pay for them. The Facility Manager has two options. He or she cannot do the mechanical work and let the complaints happen. If they do this too much, the entire mechanical system could become out of balance. The Facility Manager could do the mechanical work, but with facility funds. If this happens too often, the facility budget will not be able to pay for facility maintenance. Either option is a problem for the facility. The likely solution is for a senior manager to direct the unit or office manager to pay for the additional mechanical work.
- The unit manager wants to double up people in the current office space. She has requested new phone and electrical outlets for each new workstation. The increase in workstations will directly affect the amount of electricity needed for the new outlets. An electrical assessment might indicate that new circuit breakers should be added to provide enough additional electricity. Without this, the offices will probably suffer from circuit breakers tripping and having to be reset numerous times. Complaints will likely result from the inconvenience of the circuit breakers popping. The overheating of electrical

components is also possible, creating a fire hazard. The unit or office manager is apprised of the additional costs and refuses to pay for them. The Facility Manager has two options. He or she can do the electrical work and let the complaints happen. If they do this too much, the entire electrical system could be compromised and dangerous. The Facility Manager could do the electrical work, but with facility funds. If they do this too much, the facility budget will not be able to pay for facility maintenance. Either option is a problem for the facility. The likely solution is for a senior manager to direct the unit or office manager to pay for the additional mechanical work.

I worked in one facility where the decision was made not to require unit or office managers to pay for the mechanical, electrical, and plumbing systems required for alterations and renovations. After a few years of the Facility Manager paying for those changes or not doing them left the entire facility with antiquated mechanical, electrical, and plumbing systems that required millions of dollars to bring into compliance. The responsible thing to do is to make the unit or office manager pay for changes that are caused by their alterations or renovations.

## Cyclical Painting and Carpet Replacement

The operations and maintenance side of facility management is responsible for maintaining the facility. However, the space management side has the assets to provide cyclical painting and carpet replacement. There should be a decision made by the Facility Manager if the alterations and space management side will do this work. The money will be transferred to pay for the work. Paint is usually done at three to five year intervals. Carpet should be replaced every five to seven years. The cost of moving people, equipment, and furniture out of the way to perform the work is paid by the Facility Manager. This work must be included in contracts established for painting and carpet work. The actual cost of the paint and carpet is also paid by the Facility Manager and is included in the rent paid by unit and office managers.

## Summary

The customer is going to require alterations and renovations of space within the facility. The Facility Manager and their staff has an obligation to make sure that any work done is done properly and can be maintained. Each project is different, but as stated earlier, the size and complexity of the project determine how long it will take. The standard level of service for alterations should be:

- If the alterations project costs range from $2,000-$25,000, the Facility Office should complete the project with construction in 15-30 days.
- If the alterations project costs range from $25,000-$100,000, the Facility Office should complete the project with construction in 30-60 days.
- If the alterations project costs are greater than $100,000, the Facility Office should complete the project with construction in 60-90 days.

The Facility Specialist ensures that all contracted work is done correctly, and that proper payment is made. It is essential that all warranties are received and provided to personnel within the Facility Management Office, which will ensure any warranty work that is needed is done.

# Chapter 6 – Personnel and Equipment Moves

## Introduction

Another function of managing space is planning and conducting orderly moves of personnel and equipment with the least disruption to the employees. The prime consideration in developing the plan is the safety and health of the affected employees. The Facility Specialists are responsible for formulating and implementing a system that tracks the moves of personnel and equipment and coordinates on site the move of personnel and equipment. This work is done by a contractor that provides the workers, trucks, and equipment needed to move various furniture, equipment, and boxes. This work is normally done after hours and on weekends due to the noise and distraction. There is also the issue of freight elevators. Facilities usually only have one or two freight elevators that are used to move furniture, equipment, and boxes. This can slow down the process. Do not use passenger elevators to move furniture, equipment, and boxes. Doing so often damages the interior of the cab.

## Process

This process needs to include the specifics of the unit or office involved in the move along with the "to" and "from" locations. The Space Planner must also determine if this move includes alterations or renovations. If yes, the Space Planner will gather supporting documentation, the drawing number, and the procurement request number.

The whole process begins with the number of employees identified to be moved. The unit must determine if this is a reshuffle or consolidation. In a reshuffle, will additional space be needed or excess space vacated? In a consolidation, will excess space be vacated? In either case, the Space Planner needs to determine if swing space is needed. Swing space is extra space for the employees to be moved to while construction or furniture installation is being done. If swing space is needed, the manager of the unit or office should first look for vacant space within the unit or office. If no space is available in the unit or office the Facility Manager will try to provide space. The alterations and renovations are done as outlined in the previous chapter.

Moves that do not involve altering or renovating the workspace are necessary when relocating employees due to hiring actions or reorganization of unit personnel. These are traditional furniture, equipment and box moves. They involve identifying who is moving where and what is being taken with them. The moves are scheduled so that the employees and their items are moved at the same time, which is most efficient. It is standard practice to move some furniture in the hallway or corridor to start the process. Items moved into the hallway should be temporary and must not impede on the requirements for egress in the Life Safety Code.

Once information is gathered, a pre-move survey is completed. This involves a walkthrough of the actual employee workplace. What furniture and equipment is to be moved before construction or post occupancy? The employees of the unit or office must box up files and personal items so that furniture is not moved with contents in drawers.

There are two kinds of furniture. Case goods are the typical desks, credenzas, bookcases, file cabinets, and chairs. These are moved with little or no dismantling. On the other hand, systems furniture must be taken apart or knocked down and put back together in the new location. Seldom is systems furniture put back in the same configuration that it was in previously. Systems furniture must be designed and spare parts purchased before the systems furniture can be moved. A furniture designer and installer will need to be contracted to do this work. Spare parts must also be available to put the furniture into its new configuration. It is very hard and often impossible for spare parts to be purchased for old and outdated systems furniture. In this case, a designer can tell you what a reinstalled system would look like with no spare parts used.

Cyclical painting or carpet replacement requires furniture and equipment to be moved out of the offices so the work can be done. This is not considered an alteration or renovation. The furniture and equipment are moved into the hallway while the work is done. This process requires that all work be done for a normal move. Systems furniture will have to be knocked down and reinstalled. A redesign is not necessary if the original design is available.

**Information Technology and Telephony Moves**

The Space Planner must coordinate with the Information Technology and telephony representatives to have personal computers and telephones moved and reinstalled after the moves. It is best to do this coordination and have these representatives along on the walkthrough so they can see what work needs to be done. It is best not to have the facility moving contractor do anything with electronic equipment or telephones.

## Customer Responsibility

The customer must pay for the labor and supplies needed to move their furniture and boxes. It is unreasonable to expect the movers to pack and unpack boxes as part of the move. The drawers of furniture must have all items removed and boxed before the movers arrive. Customers will ask that file cabinets be moved with contents in the drawers. This is a recipe for disaster. During moves, the file cabinets will break apart if the contents are left inside. Furthermore, the filing cabinets with contents are much too heavy to be moved by traditional movers. There may also be occasions that a mechanical filing system will need to be moved. These are the large cabinets on tracks that move. These are hefty when filled, and before moving them, the new location must be tested to ensure that the floor will hold the weight.

As noted earlier, most moves are done after hours and on weekends. This often means overtime must be paid for the workers. This must be included in the cost estimate and budget for the move.

The supervisor of the workers being moved is responsible for those employees. The supervisor must ensure they put their items in boxes and remove files from cabinets. Employees often take moving poorly. I have seen grown men cry because they did not want to move. I have also seen many people stay home sick during moves, perhaps thinking this will help. It is not the Facility Management Office staff's responsibility to deal with these people. This situation can cause delays and cost overruns if not controlled.

## Swing Space

Many times the work and moves cannot be done when the employees are working in their office. The best option is for the supervisor to have the employees work from home. The second option is for the employees to be moved to other vacant space within the office. If this is not available, the third option is for the Space Manager to locate vacant space elsewhere in the facility that can be used temporarily by the employees. The reason I raise this again here is that this would double the work involved in the move and I do not recommend it. The people and equipment would be moved to swing space and then returned to their new space once it is available again.

## Summary

Many take the responsibility for planning and conducting orderly moves of personnel and equipment to be a side issue. It is not. This work is instrumental to successful completion of so many other aspects other facility work. Do not take this work for granted. Consider hiring a contractor that specializes in packing and moving services. With that contract should be a provision for overtime and weekend work. This will eliminate changes mid-contract. I cannot stress enough that the customer has a significant role in moving. Only if they fulfill their role can moves happen successfully and efficiently.

# Summary

My goal in this book was to share with you the valuable lessons I learned while working in Facility Management. My experience is limited to the Space Management and Alterations of Facility Management. I identified the two customers of the space management staff. First is the person who has space and wants it changed or wants more or less space and second the facility operations staff. It is one thing to meet the user's needs, but quite another to meet the building maintainer's needs. Space Managers must build out space that can be maintained. This means that sometimes the customer cannot have exactly what he or she wants because of the issues involved with the building systems and maintenance effort. The space management staff must walk a fine line to get both of these customers what they want.

The individual in charge of space management is usually referred to as the Space Manager or Space and Alterations Manager. This individual is usually a senior facility management professional. Space management division staff are facility management professionals, interior designers, AutoCAD drafters, architects, and engineers. These employees are broken down into two groups of Space Planners and Facility Specialists. Most Facility Specialists work in the operations and maintenance side of facility management, but a few work on the space and alterations team.

This is a thankless job. Everyone seems to want more, now, and at less cost. I cannot stress enough that this job involves compromise and the ability to negotiate with stakeholders to get the best outcome for everyone. The job also involves a lot of office politics e.g., people who think they are important and want more space with expensive furniture. This is not possible for everyone who thinks they are important. It requires working through the politics to get people what they need. I hope you can grow and add value to stakeholders you work with. I found this work to be fulfilling. I also hope the information in this book is helpful and usable.

For more information on this and other topics, please visit the bibliography in the next section.

#####

If you would like to help other readers out, please leave a review of this book on Amazon.com. Your rating and review will help them decide to read or not read this book.

##### 

From my other books, I recommend Management Principles for Safety and Occupational Health Managers. You can see it at the following URL https://www.amazon.com/gp/product/B01IVX5UTS/ref=dbs_a_def_rw t_bibl_vppi_i14.

# Bibliography

*3 Most Common Commercial Lease Types.* (n.d.) Retrieved on September 28, 2015 from http://www.citygrouprealty.com/Commercial%20Lease%20Type%20Training.pdf.

Addi, Gretchen and Jacqueline Lytle. *Space Planning, The Architect's Handbook of Professional Practice,* 13th edition, Washington, USA, 2000.

Cotts, David G., Kathy O. Roper, and Richard Pryor. *The Facility Management Handbook, third edition.* American Management Association, Atlanta, USA, 2010.

*Facility Management Glossary.* International Facility Management Association, Houston, USA, 2013.

Karlen, Mark. *Space Planning Basics.* John Wiley & Sons, New York, USA, 1993.

*Office Furniture Terms (Definition).* Retrieved on September 28, 2015 from http://www.primeoe.com/Services/21.

*Project Management Flashcards.* ProPros Flashcards Maker. Retrieved on September 28, 2015, from http://www.proprofs.com/flashcards/story.php?title=project-management_23.

*The Gross Lease in Commercial Real Estate.* Retrieved on September 28, 2015, from http://realestate.about.com/od/commercialrealestat1/qt/gross_lease.htm.

*Types of Leases in Commercial Real Estate.* Retrieved on September 28, 2015, from http://realestate.about.com/od/commercialrealestat1/tp/lease_types.htm.

*What Does "Net Lease" Mean?* About.com Money. Retrieved on September 28, 2015, from http://womeninbusiness.about.com/od/commercialleasingterms/g/net-lease.htm.

# References

*Finance and Business Manual*, Version 2.0, International Facility Management Association, Houston, USA, 2013.

*Leadership and Strategy Manual*, Version 3.0, International Facility Management Association, Houston, USA, 2014.

*Operations and Maintenance Manual*, Version 2.0, International Facility Management Association, Houston, USA, 2013.

*Predicting Outcomes of Investments in Maintenance and Repair of Federal Facilities*, National Academies Press, Washington, DC, 2012.

*Project Management Manual*, Version 3.0, International Facility Management Association, Houston, USA, 2014.

*Space Management Standards*, Department of Energy, May 22, 2013.

# Glossary of Commonly Used Terms

**Alterations** – The altering of space to accommodate changes in the work methods.

**As-Built** – Original set of facility drawings that include architectural, mechanical, electrical, plumbing, fire protection, and communications equipment and system.

**Building Gross Area** – is the largest area of building measurement that includes gross interior area, voids and interstitial space, and exterior wall thickness.

**Case Goods** - Office furniture that is usually pre-assembled in factories, generally not modular, and includes such types as desks, credenzas, and file cabinets (Office, 2015).

**CAD (also CAD DRAWING)** - "Computer Aided Design" -A computer program that is used to draw blueprints and floor plans of office furniture cubicles, and offices.

**CD** – Acronym for Construction Documents or a set of drawings that can be used to contract and construct a facility.

**Centralized Space Management** – Method of managing space where the Facility Manager centrally manages all space.

**Chargeback System** – A financial method of charging customers for above standard services units and offices receive.

**Cubicles** - A semi-enclosed work area built from office panels, normally containing work surfaces, storage units, provisions for electric, telephone and data service etc., - cubicles can be constructed in a broad range of sizes, heights, and configurations (Office, 2015).

**Decentralized Space Management** – Method of managing space where space is assigned to and managed by unit or office managers.

**Double Net Lease** - A type of net lease in which the tenant pays all or part of taxes and insurance associated with the use of the property. These fees are paid in addition to monthly rent for use of the actual space.

**Ergonomics** – is the science of designing the workplace, equipment, office furniture, and office furniture accessories to meet the needs of workers' safety, comfort, well-being, and productivity (Office, 2015).

**Hoteling** – The practice of not providing an assigned space for employees to work, often intended for employees that telework. Employees reserve space for use during the time they are working in the facility.

**Interior Gross Area** – The largest measurement of building interior space that includes occupant area, occupant storage, building and floor amenity area, building and floor service areas, base building circulation, major vertical penetrations, and interior parking.

**Key Plans** – A set of floor plans that show where each office or unit is located within a facility.

**Knock Down** - To dismantle office furniture or cubicles, usually for the purpose of moving or reconfiguring it.

**Lease** – A contractual agreement between a lessor and lessee to rent space.

**Systems Office Furniture** - Refers to office furniture, typically cubicles or workstations that are flexible as far as installing, making changes, moving, and rearranging.

**Power Pole** – is a vertical metal pole that brings the building's electrical service from above a dropped ceiling to connect to the electrical system of a cubicle or group of cubicles (Office, 2015).

**Power Whip** -A term used in the office furniture and electric industries to refer to a flexible connector that contains wires, and connects the building's electric service (usually in a wall) to the electric system of a cubicle or group of cubicles (Office, 2015).

**Reconfiguration** - Dismantling office cubicles and rearranging them into different sizes and shapes, using only the existing components, or with a combination of other compatible components (Office, 2015)

**Relocation -** A complete or partial office move of cubicles, and private offices.

**Renovation** - The upgrading of office space to accommodate changes in work methods.

**Rentable Square Feet** – The measurement of building an interior that you pay rent on that includes occupant area, occupant storage, building and floor service areas, building and floor amenity areas, and base building circulation.

**Single Net Lease** - A type of commercial real estate lease in which the tenant pays for their space, as well as for part of all "usual costs" that the landlord pays (What, 2015).

**Swing Space** – Space that is used while alterations and renovations are being conducted.

**Standard Services** – The service a unit or office receives for the basic rent payment for assigned facility space regarding rentable Sq. Ft.

**Teleworking** – The practice of allowing employees to work from home using a computer connected to the net along with the telephone with the goal of reducing or eliminating worker commute.

**Test Fits** – A drawing of space that is done to scale and shows how furniture and equipment will fit.

**Triple Net Lease** - A type of net lease in which the tenant pays all the "usual costs."

**Usual Costs** - Expenses incorporated into net leases may include taxes, utilities, janitorial services, property insurance, property management fees, sewer, water, and trash collection. Net leases almost always favor the landlord and should be negotiated to include caps, or, the maximum amount a landlord can increase fees each year (What, 2015).

**Usable Square Feet** – the smallest measurement of building an interior that includes the portion of a floor or building used to house personnel, equipment, fixtures, furniture, supplies, goods or merchandise, and secondary circulation (Project, 2015).

## About the Author

After a successful career as a Federal Employee that included over twenty years in safety and occupational health. I started writing part-time. My published work includes the peer-reviewed book Basic Safety Administration: A Handbook for the New Safety Specialist in its second edition. I also authored two editions of the peer-reviewed chapter, Safety Training and Documentation Principles published in the bestselling, Safety Professional Handbook, and the Safety Professional Handbook Management Applications, both edited by Joel Haight, Ph.D., CSP. I co-authored the peer-reviewed chapter Safety Training with Christine Fiori, Ph.D., PE, published in the bestselling Construction Safety Management and Engineering, second edition edited by Darryl C. Hill, Ph.D., CSP. The American Society of Safety Professionals Traditionally published my book and chapters.

I self-published another eleven books using Kindle Direct Publishing. Seven of these books are available in paperback and Kindle formats. Four of those books are available only in Kindle format. I have authored over fifty articles in various publications on safety and occupational health and project management. I have earned several writing awards for my non-fiction work and one for my fiction work. I have self-published two novels, A Walk Among the Dead and my most recent Mystery at Devil's Elbow.

I am an Emeritus Professional Member of the American Society of Safety Professionals. I was selected as the Safety Professional of the Year for the Northern Virginia Chapter of this Society. I am also a member of the Non-Fiction Writers Association. I held the Certified Safety Professional (CSP) designation for ten years. I also earned master's degrees from National-Louis University and Webster University.